I047040b

Astronomy and Humankind

A New Interpretation of Certain

Celestial and Terrestrial Phenomena,

and of their Significance for Human

Life on Earth and *Elsewhere*

By

James H. Pax

ISBN: 0-7596-9879-1 (softcover)
ISBN: 0-7596-9878-3 (ebook)

This book is printed on acid free paper.

1stBooks - rev. 03/13/02

This work is dedicated to our recent

hard working, long suffering Parents

Antonette Marie (Nordenbrock)

and

Jacob George Pax

Table of Contents

1. Introduction

The subject of this slim volume is not the whole of astronomical science, but rather astronomy and its relation to life on Earth. The aim of the author is not to present another textbook of astronomy. His aim is to set before the reader certain astronomical ideas, some of them new, that are of special interest because of their bearing upon life on earth, and their relevance for understanding man's place in the universe.

Thus the aim of the author is at once modest, in that it considers only a portion of the field of astronomy, and yet daring, in that it offers a radically new interpretation of some of the phenomena under consideration. Because some explanations offered are new, the book, in spite of its brevity, might well present a gigantic challenge to the reader. The text seeks nothing less than to prepare the ground for a revolutionary conception of the universe in which we live. Hence it calls also for a significant reconsideration of the common understanding of our place in the universe.

The reader is asked to set aside for the time being his or her commitment to conventional wisdom and to read the matter here presented with an unprejudiced mind. Because the overall picture here presented is in many ways new, it is not likely that all of the assertions that are made will appear immediately evident. An endeavor that seeks to engender a radically new understanding must inevitably begin on a tentative note. Not all of the elements that go into the picture will be immediately obvious. Indeed, the significance of the individual parts of the picture might not become accessible except in the consideration of the whole. The hope of the author is that those elements of the conception presented that do not appear immediately credible will become clear and worthy of serious consideration from a careful study of the whole.

Even with the whole picture laid out, however, the new conception cannot gain acceptance unless the reader is able to approach the subject with an open mind free of excessive adherence to former understandings.

Moreover, the author realizes that much work remains to be done in order to make the picture complete and scientifically satisfactory. And although he takes pains to present his ideas and convictions with clarity, he realizes that the newness of the vision presented will bring forth well-meaning opposition. Such serious opposition is welcomed. The author's hope is not that every piece or element of his vision and his reasoning will be seen as satisfactory, but that the whole of the picture he presents will be seen to be worthy of serious study. He will hold himself well rewarded if his work sets the stage for and inspires further study.

Within a new conception of the universe and man's place in the cosmos, not every idea presented need indeed be acceptable. What is important is that the whole congeries of the ideas presented work together to reinforce each other and thus to entice the reader to move into the challenge that has gripped the writer for the past ten years.

The author gladly acknowledges his dependence upon the work of the many careful observers and recorders who have worked in the field. He expresses his deep respect and gratitude for these worthy forebears. His effort and desire is simply to share his further reading of the evidence that has been written in the structures and the movements of the universe by its Creator. The evidence is before our eyes. We need only to read it with eyes that are newly attentive to the work of the Maker.

The author gratefully acknowledges the editorial assistance of his brother Clyde.

2. Our Solar System

When we here speak of our solar system we mean that part of the astronomical universe that is centered upon and revolves about the star that we normally refer to as the Sun. Our solar system has a definite size and structure; and it has dynamism. Mass, distance, velocity and kinetic energy are all important for understanding our solar system.

The Sun is about 100 million miles distant from the Earth, is approximately 800,000 miles in diameter, and moves toward the North Star, which is 80 light years distant. By contrast, the light from the Sun reaches the Earth in about 8 minutes.

To date, nine planets that circle the Sun have been discovered. Mercury, Venus, Earth and Mars are the "rocky" planets; Jupiter, Saturn, Uranus and Neptune are gaseous planets. Pluto, only half the diameter of our moon can be called the rogue planet. It rotates on its axis while lying pretty much on its side, and it flies backward; it is only two percent as large as Earth. Pluto does, however, have a moon of its own, and as we shall see, this means that Pluto is likely subject to internal heating. Hence it could support vegetation that in turn produces oxygen; therefore it could possibly support life similar to that found on Earth. Pluto intercepts the orbit of Neptune and therefore will likely, some billions of years in the future, impact Neptune.

While Earth is about 100 million miles from the Sun, Mercury is about one third that far. Venus is three fourths as far, Mars (the sister planet of Earth) is one and half times as far, Jupiter 5 times as far, Saturn 10 times, Uranus 20 times, Neptune 30 times and Pluto 30 to 50 times as far from the Sun. Pluto's orbit is 243 years.

The Earth and all the planets more remote from the Sun fly at about the same speed, roughly 600 million a year (66,000 miles per hour). Mercury and Venus, already on their way to falling into the Sun, travel a little faster. History mentions another planet called Vulcan that is said to have entered the Sun a thousand years ago.

Most of the planets circling the Sun, our Earth included, have satellites of their own, captured in orbit by the parent planet. These satellites are often called moons, even though, strictly speaking, this is the name of the satellite circling Earth. As will be argued more fully later, the moon of the Earth is absolutely necessary for life on Earth. Without the subsurface heat produced by the constant massage of the Earth by the tidal force of the moon as it makes its daily cycle around its captor, Earth would be too cold to support life as we now know it. The plane of moon's orbit is shifted by the tidal tug of Jupiter. Hence the string of the Hawaiian Islands, ever bigger.

Astronomers have discovered twenty-two moons circling Jupiter, four of them as large as that of Earth. Temperatures on Jupiter, however, are too hot for life, as we know it. Saturn, the planet nearest to Jupiter, has a moon called Titan that is about the size of Earth's moon. Neptune also has a moon, named Triton. The day on Titan lasts about 16 hours and on Triton about 6 hours. In the far distant past the day on Earth was shorter than it is at present, but probably not much shorter at the time of the beginning of life on Earth, which is thought to be about *one-half billion* years ago. If all of the estimated 400 billion stars in our galaxy have planets with moons found as frequently as they are found in our system, then there may be over a trillion sites, just in our galaxy, with a climate that is suitable to support life. I believe life travels between the planets as DNA, shed in the comet like tail and re-cultured on the receiving planet, such as ours. A trillion sites would figure one per 400 million suns, or one per 3.6 billion planets.

Mars has two extremely small moons at present, but at one time must have had a big moon like Earth's moon. Evidence for this larger moon is found in the presence of the cooled off volcanoes on Mars. When Mars lost is large moon, (I suggest in disintegration caused by some collision) the outermost layer of the planet shrank as it cooled, squeezing any molten rock from underneath the crust to the surface. First there was a bulge, and then the emerging volcano formed a mountain larger than any mountain on Earth. The larger of the two moons currently cycling Mars is a rock, shaped something like an elongated potato, measuring about 150 miles in length. These two small moons are fragments of the planet's earlier moon. They are too small to give any reason for us to think that they can produce enough heat on Mars to enable the planet to sustain life on its surface. In orbits roughly between Mars and Jupiter there exists 30 or 40 thousand orbiting fragments of busted planets. These are evidence of two planets that met.

Asteroids coming to earth over the ages are some of these, and are of the composition corresponding to the core and mantle of these lost planets. Even the iron-nickel meteorites from the core show extreme weathering, *a calendar* to the incident.

When we note the passage of days and nights, or marvel at the repeated phases of the moon, and the patient and predictable movements of the stars in the heaven, we may be tempted, as were the ancients, to think that all the movements within the universe are regular and periodic motions. This is far from the true state of affairs. Regular, periodic, motion is, of course, generously present in the universe and is of tremendous importance and interest. But many irregular motions, some of them hugely disruptive of the shape and structure of the entire system accompany this regular motion. In addition to collisions between masses, there are explosions within masses and exchange of mass and energy among the various larger bodies. There is also a slowing down of the periodic, circular motion that characterizes both the stars and the bodies captive to the stars. *The "no free lunch" principle.* A captive such as the planets to our Sun, eventually join the captor, due to friction in space. The law of conservation of momentum is that when two bodies join, the result is a compromise.

One significant example of this slowing down of velocity is evidenced in the rotational *freeze* that has gripped our moon. At present the moon shows the same face to the Earth at all times. The grip of Earth on its captive is so strong that the moon can no longer turn on its own axis. The big lava flows, however, on the surface of the moon can only have been caused by friction of the Earth's tractor beam at a time when the moon had an angular velocity.

With this brief introduction to Earth's larger companions in the solar system, it is time to turn our attention to the some of the effects that living in this company has had upon Earth. In doing so we will have to take note not only of the larger bodies, but will also have to read the evidence left by the smaller bodies, even by very small bodies, that is *sub-atomic* fragments. And we will need to attend to the forces that hold this large family together. Above all, we need to accustom our minds to read all of this evidence within a framework of immense distances, great velocities and time frames far beyond our normal experience.

19

Nonetheless, the evidence left by the Maker who made both the cosmos and *our minds* assures us that we need only read carefully what is before us. Man *assigns* a beginning and end to things he doesn't understand, because man suffers a beginning and end. I assign this to the EGO FACTOR.

3. The Recent Origin of our Earth

We don't know the age of the Earth, but we do know that it is a comparatively recent arrival in the solar system to which it belongs. Earth is a captive of the Sun and goes where the Sun goes, and at the *same speed* as the Sun. Was the original Earth, of perhaps 20 billion years ago, a chunk of some larger body destroyed in a collision; and did Earth then later accumulate its mantle from the scattered debris of interstellar space as it followed the Sun during long eons? The Hubbell Telescope has spotted a light source *60 billion* years old.

The longer a planet is in orbit, the more circular the orbit becomes.. The Earth's orbit is almost circular.

The core of the earth is known to consist of a metallic alloy of iron with a ten- percent mixture of nickel. Such an alloy has a specific gravity of 8. If the average specific gravity of Earth's mantle of rock and clay is about 3, and the earth displays an average gravity of about 5.5, then the mantle must have an average thickness of 400 miles. The Earth has the greatest gravity of the 9 planets. The weight exerted by such a mantle is calculated to create a pressure on the inner core of 12 million

million pounds per square inch. Added to the pressure from the rocky mantle of the Earth there are, at present, 14.7 pounds of air pressure pressing down on each square foot of the Earth's surface;

This combined pressure exerted on the center core of the Earth tells us that the core is fluid no matter what the temperature is "down" there. The best iron alloy that we know has a strength of about 300 thousand pounds per square inch. The compression on the inner core of Earth is many times greater than the burden that the finest metal alloy can bear on the surface of the Earth. Earth's core is fluid by virtue of *sheer force*. So earth assumed a spherical shape!

I believe it is misleading to call the mantle of the Earth a "crust", since this term implies that the mantle was molten, and that the inside of the Earth is still molten. But this is not the case. The core of Earth is fluid because the pressure exerted on it is many times the pressure that would overburden the resistance of the finest tempered steel. The core of the Earth <u>acts</u> as though it were <u>liquid.</u> *It's temperature is immaterial.*

Was the original Earth a chunk of iron alloy that became covered by the present mantle acquired during the Earth's travels over billions of years? Perhaps. It could also be the case that the denser iron and nickel alloy below the mantle came to exist as one result of a prolonged natural *transmutation* of the elements, akin to but different from the transformation that takes place within the Sun. As we now understand the matter, the Sun appears to be a huge moleculovoir, as it were, that devours whatever gets in its path and then sends forth the transmuted *more simple* material as radiation, and as a *trail.* These trails are extremely important to us, as the Earth cuts through them. Could a related but vastly different transmutation take place in the smaller bodies in space such as the Earth? Chemistry and physics *overlap.*

If natural composites do tend to sort themselves out according to their density, then most of the gold in the composition of the Earth must have gone to the central core eons ago. Humanity will always have that gold safe in the most permanent of banks. On the same supposition, namely, that composites sort themselves out according to density, the most logical place to dig for heavy metals is beneath the area about the South Pole, since the material of lighter weight and density continually settles much more abundantly, though not exclusively, about the North Pole, leaving the South pole with the old, old stuff.

Among the *materials* striking the Earth is light from the Sun. Although we do not often advert to the fact, sunlight has weight. The sunlight that has struck the Earth in the last 5 billion years has been estimated to have a weight of 10 to the 17th kilograms. A portion of this sunlight has bounces off the Earth again. The part that has remained with the Earth, however, must have left a significant amount. Even today the composition of the mantle of Earth corresponds to the spectrum of sunlight!

The sunlight is also, of course, an important source of heat, absolutely necessary for the kind of life that exists upon Earth. We will see, however, that sunlight is only one of the necessary sources of heat that enable the Earth to sustain life. As important as is sunlight, and as important as is the solid matter that now forms the continents of the Earth, still other materials gathered on the Earth's prolonged trek through space are equally necessary for life. Most notable among these other gifts are hydrogen ions (protons) gathered from interstellar space, which through another transformation give us both water and ozone. Note also that the magnetic field of the earth is continuously generated, hence the fluctuation Electrons are much more plentiful that we are led to believe.

4. The Origin of our Continents

How long ago and in what manner the Earth became a companion to the Sun is not known. Ever since it became bound to the gravitational field of the Sun, however, during billions of years, our Earth has traveled as a captive, going where the Sun goes. During this long trek, the spin of the Earth on its own axis kept the North Pole first (less 23.5 degrees) in the line of travel, like a good football pass. At what speed has the Earth been traveling through interstellar space? The Earth travels in its orbit around the sun at 66,000 miles per hour. (About 600 million miles in one year.) But the Sun follows an orbit of its own, a much bigger orbit related to the (yet unknown) anchor star to which it is captive, and the Sun travels at a much greater velocity than does the Earth around its anchor star, I estimate the speed of the Sun at about 200,000 miles per hour! For billions of years at 200,000 miles per hour the forward face of the Earth has been sweeping interstellar space. A spot on the Arctic circle is always fore most establishing a ring of material, at one time the Rockies and Andes.

Just as the nose cone of a modern space rocket strikes whatever debris it meets in space, so also the forward nose of the Earth, that is, the area surrounding the North Pole, strikes whatever lies in the path of the Earth. For billions of years, and at tremendous velocity, the Earth has been sweeping through space and locking unto itself whatever bodies, great and small, accumulated onto its forward face.. Bodies did not simply fall unto the Earth. Rather, the Earth must be seen as an aggressor. over taking the bodies. In the course of its aggressive and continual travel, the earth scooped up a huge mass of matter, locating it principally on the Arctic circle. After many eons, this ever-growing deposit formed itself in a ring all around the forward nose of the globe, that is, in the region we call the Arctic Circle. A spot on the Arctic circle is always most forward, and each spot takes it's turn, once a day, to be the most forward.

As the Earth traveled though space, it also spun on its own north-south axis. This spin exerted a centrifugal force on the material gathered on the surface at the forward nose of the Earth. Driven by the centrifugal force caused by the spin of the Earth, this collected material slid and was pushed south retaining the shape of a ring because *all directions from the North Pole are south.* Perhaps the debris moved only a short distance in thousands of years. But as the centrifugal force shoved the matter further from the region of the Pole, the ring grew both in volume and in distance from its original position. As still more material accumulated, the push southward increased. The continual push southward was at times halted by the frictional resistance to the movement of this increasingly immense mass of materials, perhaps *miles* deep. I n South Africa there is a gold mine two miles deep showing sedimentary rock! This means that accumulation at the Arctic circle was at least 2 miles deep at times before a shift! The coupling of the centrifugal force against the frictional resistance, both abetted by the continued collection of ever more matter, caused the mass of matter to *bunch up*. The thickest part of this accumulation formed a ring all around the Earth, approximately at the latitude of the Arctic Circle. This ring rebuilt many times. Eventually, after how long a time we don't know, these combined forces formed a stabilized ring of young mountains all around the North Pole at the latitude of Alaska in North America. More than fifty percent of the last ring is in place today. The evidence is still there, written in stone, for us to read. Because dinosaur fossils tend to be just east of the rocky mountain backbone of the western hemisphere, I believe the critters lived on the shore of Lake Steven until about 70 to 140 million years ago. Gravel and petroleum deposits (I believe) also happened about 100 % on these shores. Petroleum is the *feces* of marine creatures, we should be looking for the DNA content, and pathogens A river doesn't have the force and action to produce "river" gravel, but a sea shore does. Consider the portion of each gravel pebble that is missing, it is our soil. Gravel and crude petroleum deposits largely occur together. As mentioned otherwise in this document. An oil well is akin to putting a soda straw into a privy.

It is not known how long this original ring of mountains and the sea to the north of the ring, endured. Eventually, however, the forces pushing from behind the mountains did cause a break-up of the ring and a rearrangement of the accumulated landmass. The span of the determined ages of the dinosaur fossils in the North American Dakotas and in South Americas tattle on the span of *one* time when the mountainous ring of the Steven Sea held fast. *One tiny chip to our puzzle.* This ring was constantly rebuilt. The break-up and rearrangement of this original ring of mountains accounts for what is now the string of mountains that form the backbone of North and South America. This string of mountains is roughly equal in length to the length of the Arctic circle.. Days are shorter now than in the past.

An incidental but important result of the collection of this material from space is that the added mass of the Earth slowed down the velocity of the Earth in its axial rotation and orbit about the Sun. Because of its decreased velocity, and increased mass, the Earth moved into an orbit closer to the Sun. This greater proximity to the Sun increased the amount of sunlight that Earth received from the Sun, and helped to raise the temperature on the surface of the Earth to a point more favorable to support life. The light from the Sun is, as we shall argue later, only one of three major sources of heat on and within the Earth. The plane of the orbit of the moon with respect to features on the Earth changes over time. From time to time the moon has a polar orbit. And a times moon's orbit is just backwards to present! This is caused by the tug of Jupiter, a

thousand times larger than Earth { in volume}
and a thousand times smaller than our sun.
(The Earth is about 8000 miles in diameter,
Jupiter 80,000 and the sun 8 hundred thousand)
Volume is a function of the cube of one
dimension. the less sometimes our QED! The
result of these tugs is called perturbation. But
Earth has to live with a nuisance like a *polar*
orbit. During a period of polar orbit the moon
helps the centrifugal southward distribution of
continental mass. Note that although Jupiter
cannot steal our moon, it can change the
moon's plane of orbit. Let us now direct our
attention in more detail to the forces that made

this original group of mountains break up. We need to discover how these forces combined to move and shift the material on the surface of the globe in such a way as to create the present shape and arrangements of our continents. Again the evidence is still there; we need only observe the clues that are available and draw them into a coherent picture. The new knowledge to humanity of the two separate forces which can move the mass and make mountains, make it easy.

In the early stages of the Earth's existence, its surface was devoid of water And if there were an atmosphere, it was different, the amount of atmosphere is always changing. Already then, however, the Earth's work of gathering mass was underway. And the spin of the Earth, overcoming the frictional resistance offered by the surface (due to Earth's own gravitational grip on the stored-up matter), was thrusting the increasing mass southward away from the nose of the planet. The delicate interacting of the centrifugal and the gravitational forces is itself a clue left by the Maker for our instruction. Moreover, this particular interaction is one of many such marvels. We have already called attention to the marvel of the companionship between Sun and Earth, and to the whirling spin of the Earth as both Sun and Earth speed toward the North Star. All of this is evidence to be read for our enlightenment. So constant and universal is the evidence that it can be said truthfully that without ken of God there is no human knowledge. As the collected mass increased in volume the thrust southward increased, and the evidence present to us today is that this centrifugal force did in fact overcome the frictional resistance that tried to hold the matter in place. As the mass moved, it moved southward roughly in a ring away from the nose of the planet. Please keep in mind the

43

assistance of the ice age blanket, two miles thick. An ice age occurs about every 60,000 years. The present interlude should last another 50,000 years, during which the oceans will continue to rise, flooding most of the present cities. During the past 10,000 years there was a 200 foot rise in the ocean level. Remnants of past civilizations are found in the Mediterranean along the coast of Africa, Stone walls and stone pavements. Man thinks he's schmardt when he alludes to the lost civilization of Atlantis, *COME TO THE PARTY,* there were thousands of human communities, covered and destroyed by the rising waters, and forgotten. There exists today, a river gorge between Scotland and Europe covered by many feet of water. This river was made when the oceans were below that level. If one wishes to discover lost cities look under the deposits at the major river deltas, where they are preserved by the ancient deposits of the river's mouth. Core drilling will show us just where the river was at any specific time in the past.

With the arrival of water on Earth's surface the forces affecting the movement of the landmass were dramatically altered. (Vegetation makes the oxygen, which combines with hydrogen from space to make water, as stated) Before plant life, no oxygen. Tell me how long it took. Not only did the frictional resistance offered by the surface of the Earth decrease with the lubrication by the water, but the centrifugal force from the rotation of the Earth increased as the weight of the water (*ice*) was added to the mass. Eventually the huge basin created by the mountains encircling the North Pole filled with water. This huge body of water, about 4000 miles in diameter, and which I herewith christen the Steven Sea, must at some time have been a "Garden of Eden". As in the

Ancient Bible. (Think of this sea as a 4000 mile diameter petri dish, receiving and culturing DNA constantly received from interstellar space during billions of years) The large petroleum deposits discovered under the surface in this region, as well as fossil remains, are evidence that the shores of this sea were once witness to an abundance of plant and animal life. This "Garden of Eden" was warmed not only by sunlight but also by two other sources of heat, namely the hydrogen burning over the northern pole which was directly over it and the tidal massage of the moon, both of which will be clarified in a subsequent chapter.

The water of this large sea (many times solid ice two miles thick) was constantly subject to the thrust southward caused by the spin of the Earth. And if the Earth already had its moon at that early time, the tidal force of the moon also exerted constant insidious always westward pressure on the primitive, still soft, ring of mountains holding back the wall of water. These forces combined and as more and more material was scooped up from space, the ring broke in spots time and again and released a flood. One example of such a break through the mountain ring is identifiable still today in the eastern part of the state of Washington USA. The surface there is scarified off to an reported depth of 80 feet. During repeated ice ages, the break-up of the land was furthered by the southward push of *millions* of tons of ice. This ice, like everything movable on the surface of the planet, was subject to the centrifugal force. The ice moved south in the form of powerful glaciers, typically taking large portions of the landmass (now continents) with them. Ice moved south, whether down hill or not. Normally, ice squeezes out from underneath the rigid ice surface. At a depth of more than 91 feet, ice is plastic, and *flows*.

Continental drift, that is, the movement of mammoth chucks of the rocky mantle of the earth, is no longer unfathomable once we understand the magnitude and the duration of the forces applied to these huge and seemingly unmovable parts of the Earth's surface. The shift of the landmasses from north to south was caused by centrifugal force as a result of the spin of the earth. Working in concert with this southward thrust was a westward pull caused by the tidal forces of the moon. Hence the *pivot* of North America. about Alaska. North

America is presently being rent on a line shown by the great lakes and the lakes of Canada. The finger lakes of upstate New York are rift valleys—***stretch marks***. Lake Michigan is a rift valley—filled with water. The Pacific ocean may be considered a rift valley, explaining the drop it it's floor, as shown by *truncated* sea mounts. Australia split off from South America.

Looked at from the apparent regularity of our days and nights the rotation of the Earth appears to be constant and fixed. Looked at from astronomical periods of time, however, the spin of the earth is not a thing that is constant and to be taken for granted; it is a dynamic and dying thing. There is good evidence to believe that the earth spun faster in the past. And there is the real possibility that the Earth will eventually stop spinning. The work of holding the moon in place will become sufficient to put the earth into a tidal freeze in relation to the. Earth's moon. Our moon, is already in a tidal freeze; that is to say, the same side of the moon is always toward us. Yet we know that the moon at one time in the past turned on its axis. This is confirmed by the huge lava flows on the moon, once thought to be seas, hence the designation "mares" from the Latin word for seas. These flows would not have formed with the moon in its present frozen spin.

The power of the thrust southward, away from the North Pole, caused by the rotation of the Earth increases with the size of the mass that is moving. During the ice ages, which, if we can judge by the record of the last quarter million years, amount to two thirds of the time, a giant slab of ice thought to be 10,000 feet thick, helped to push all land south. From the North Pole any and all directions are *due south*.

Whereas the spin of the earth moves things south by centrifugal force, the force of the gravitational attraction beam between the Earth and the moon is always toward the west. The moon seems to rise in the east and moves toward the west, pulling whatever will move on the Earth toward the west. Huge boulders resting on the level floor of dry lakes move to the west. What moves most readily are the oceans, and we experience this tug of the moon on the Earth as the tides. The tidal force acts, however, not only on the oceans but also on the masses of solid land surrounding the oceans. The tidal force pulls upward and westward. The necessary upward thrust of the tidal force makes the notion of subduction, that is, the idea that one continental plate is drawn underneath another an impossible notion. What takes place is <u>overduction</u>, namely, the drawing of one plate over the top of another in its westward migration. *There is no* <u>*sub*</u>*duction.*

To read the record of the work done by these two forces we need to make repeated efforts to appreciate the magnitude of the forces involved. At the equator, where the centrifugal forces reaches its apogee, the surface of the Earth travels at about 1000 miles per hour, and does so continuously. The tidal force of the moon, in its turn, is sufficient, for

55

example, to lift and lower the water in the Pacific Ocean, and to change this pacific ocean into a huge *diaphragm* day after day. These forces, one pushing southward, and the other always westward, work in consort, as well as in their own right. And they have had billions of years to complete their tasks.

From study of a topographical map we can read the evidence of the work already done by these forces. We can learn from observation, for example, that South America split off from Africa and moved west about 3000 miles, and rotated about 20 degrees since the split, but Africa has also been moving, with a little tearing—hence the 3 large rift valleys. How long this event was underway, for now we can only guess. Some astronomers have suggested that the move was completed within a period of 50 million years. We can see also that Australia has split off from South America, showing us that the Pacific is newer that the Atlantic! This split is the most likely explanation for the existence of mountains on the eastern coast of Australia. Added evidence for the supposition that South America and Australia were once united is the existence of marsupials on these two continents, and for a long time *only* on these two continents.

Possums came to North America only "recently", after North America swung down to create the Isthmus of Panama. South America is also coming north, a correction from being shoved south during an extra large ice age. Note that a sheet of ice covering almost the entire northern half of the earth would starve out the grass eating types, starving out carnivores. The dinos starved. some mammals took to the sea rather than die. Ice ages last more than a thousand years. Whales, dolphins.

We can note further that the Atlantic coast of North America once joined the north coast of Africa. Europe, including England and the Alps, have moved south and west. *And the Arabian Peninsula has inserted itself between Egypt and India*, which have congruent coastlines. (I believe that Arabia picked up it's petroleum as the excrement of marine creatures when Arabia was still located at the Arctic circle) Studying the map further we can locate areas where a landmass is shoved to the west, unto a landmass that won't move. The Indian subcontinent and Alaska appear to be so deeply anchored as to be relatively locked in position. Whether they will stay locked in position is presently beyond our knowledge. When the ocean levels are high, landmasses move more readily because of the buoyant effect of the water. <u>The buoyancy effect of high ocean levels appears to be a larger factor in easing that movement due to the tidal forces (westward) than the movement caused by the centrifugal force</u>. During the ice ages, a sheet of ice about two miles thick assists the southern movement of the land masses. A longer ice age would have thicker ice, extending further south, South America being pushed beyond the equator is an attest to that. India offers another forceful clue for reading the evidence of continental shift. There are seven or perhaps ten mountain ridges immediately to the east of India (caused by the tides) and a like number to the north of India.(caused by centrifugal force) Each mountain ridge represents a major crunch. Note that there is only one major ridge running north and south to the west of India. Both India and Alaska have (pretty much) refused to move for a long time.

Looking again at the map we see that the ocean floor is "out-gassing", allowing pent-up heat to break forth. Consequently we see ocean trenches as the ocean floor shrinks. Is it completely unreasonable to think of this release of gas indicating a transmutation of elements? We are told that the earth's core is an iron alloy with a little nickel. Such an alloy has a specific gravity of 8. And we are told that the earth has a specific gravity of 5. How thick must the mantle of the earth be, assuming its average gravity is 3? Answer: about *400 miles*. Man can physically sample his earth to a depth of about 30,000 feet. Now is the time to speculate about the interior of the Earth, using, of course, all that humanity has so painstakingly observed. My heart and my respect goes out to all who have gone before, diligently observing, recording and saving to our store of knowledge. Thank God for our accumulative culture.

5. Water and Ozone

Let us suppose that the earth made its appearance in the solar system as the result of an interstellar collision. In the long-distant past this chunk of matter was drawn into the gravitational force of the Sun. As a captive of the greater body, Earth began its long voyage through interstellar space. Note that Mars and Earth have some relationship, both have an axis of rotation that in parallel and turn at about the same speed, Mars just 2% faster than Earth. Has Earth slowed down more?

If we can judge from the present knowledge of the core of the Earth, the original chunk would have been a metallic alloy, chiefly iron but with a small percentage of nickel. Most likely the surface of the early Earth was devoid of water, perhaps for a billion years or more, and there was as yet no atmosphere surrounding the Earth. Figure that vegetation produced the oxygen in the atmosphere, as it does now. During this long period, however, Earth was already gathering matter on its frontal face, in its high velocity sweeping of the area through which it was drawn as a companion to the Sun. This matter included light from the sun, solid matter from other collisions, and very importantly, hydrogen ions (*protons)* that combined with oxygen to form both water and ozone. Hydrogen is thought to make up eighty percent of the matter in interstellar space. The best bet for a source of oxygen, then, would be bio-oxygen from very simple vegetative life, flattened onto the surface of Earth in its passage through space. Life conditions would have been harsh and the form of life would be simple, but simple vegetative life still exists in the region of the South Pole, even at eighty degrees below zero.

There could have been more than one kind of primitive life form, including perhaps bacteria.

The hydrogen ions collected on the earth's frontal surface combined with the bio-oxygen, and in the process produced water, ozone and heat. In the passage not of centuries but of millions or billions of years, this initially minute combination of elements grew into a large reaction zone located chiefly around the North Pole where the ever renewed collection of hydrogen provided ample fuel for the reaction. When we look into the northern lights, we are looking directly into this reaction zone, often some 4000 miles across and thick enough to burn all the fuel furnished in the form of small electrically positive particles. A hydrogen-oxygen fire is quite colorless and doesn't shine much,. Traces of sodium and carbon would give off a similar light to that of the northern lights. The spectrum of the northern lights will show us exactly what else is normally in interstellar space. Now we know!

The magnetic north is continuously generated by the passage of the frontal nose of the Earth. The availability of the fuel, that is, the density and location of the ionized hydrogen, causes the magnetic north to wander, and can even allow the pole to reverse its charge. Both chemistry and physics can claim a role in the explanation of the working of this reaction zone: the relatively simple chemistry that details the order and the results of the combinations of elements and the relatively simple physics that details the generation of a magnetic field.

As noted above, the reaction zone provides *heat, water* and *ozone.* All three are vastly important and indeed absolutely necessary for life, as it is now present on earth. (The ozone clears the air of methane and ammonia, both poison to man) This atmospheric system has already been fixed for perhaps 3 billion years. We should look for the spectrum of the northern lights about the nose of Pluto. To extend this thinking a little farther, into the realm of speculation, we might conclude that any heavenly body that has water, once had vegetation, to produce the oxygen component. To find life, look for evidence of the existence and operation of the reaction zone detailed in the last two paragraphs. Consistent with the presence of fire at the North Pole and its absence at the South Pole is the established fact that the average ambient temperature is consistently higher at the North than at the

South Pole. Furthermore, there is ozone about the North Pole but not at the South Pole. The hole in the ozone layer at the south pole grows and shrinks, as the amount of encountered hydrogen varies. About the North Pole there is greater precipitation than about the South Pole, There is more rainfall over the north because the water is <u>made</u> in the north. During an ice age the water simply stays at the north and is *solid* - ice.

The polarity of the earth's magnetic field <u>reverses</u>, and does so on a regular basis. (This has been noted along the mid-Atlantic ridge) The periodical character of this reversal suggests that it is tied to a cycle, that is, to the *trail* orbit of some astronomical object. like the sun. Suppose that the Earth generally passes though areas of space that are rich in hydrogen ions (protons), but that occasionally Earth passes through areas that are lean in protons, and hence a reversal of polarity in the magnetic field. There is no need to suppose, as has been suggested, that the billion trillion-ton Earth does a flip-flop. The Sun must at times take us who dwell here on Earth toward the *inside* of the galaxy, and at other times toward the *outside*, where interstellar matter is sparser.

The amount of fire over the North Pole is related to the height of mean sea level. In the past million years there was a time when the oceans of the Earth were 200 feet lower than at present. Evidence for this is the observed presence of the continental shelf just off the coasts, establishing a previous and well-defined standard sea level some 200 feet below where it is at present. The continental shelf "is there"! The evidence is there for our reading. The oceans are now rising again, and have been since the last ice age, some 10,000 years ago. If the pattern of previous ages is repeated, we can expect the sea level to rise another 150 feet or so, and then return to the level of the continental shelf. About a 60,000 year cycle. Judging by the last half million years, the present limit of our knowledge, Earth has sustained an ice age during most of the time that Earth existed. About a thousand ice ages since the demise of the Dinos.

When the fire of the northern lights abates the production of water and ozone also decreases. With less water being produced, the level of the oceans drops. The drop is made more severe as water gets tied up in ice because the heating stove in the north is burning less furiously. With less heat from the stove in the north, the ice cap grows. An ice cap 10,000 feet thick over about 20 percent of the land, which is only one third of the earth's surface, would drop the ocean level only about 100 feet. The additional drop of 100 feet can be explained by the supposition that less water is made during an ice age.

The lowering of the northern lights also causes the ozone layer to thin out. The ozoneless hole at the South Pole grows larger, perhaps extends even as far as the equator. The play of these astronomical forces has much more, by far, to do with the amount of the ozone than do the activities of the human creatures who *depend* upon the ozone cleansing of the atmosphere.

An interesting and related topic that needs further study is the question of how these atmospheric and surface changes on Earth affect the loss of matter in the mass of the earth. If Earth has a tail like Haley's Comet, the Sun, and perhaps most heavenly bodies have a tail. How much matter is in the tail and in what form is this matter present? Some Russian researchers have estimated that the earth loses 14,000 tons of atmosphere per day. A study of this matter might bring about refinements in the estimates of the change in the amount of total water on earth or of ozone

in the atmosphere.

6. Three major sources of Heat on Earth

Before we consider the chief sources of heat on Earth, it might be well to consider briefly the importance of heat. What comes to mind first of all, of course, is the absolute necessity of heat for our lives, and for all life, as we know it. We simply could not live if we were exposed to the low temperatures that generally prevail in interstellar space. So it is of some importance to understand and appreciate the sources of heat. A lot of heat must be continuously supplied.

Heat is important also because of its relation to the issue of motion. All movement of mass is equivalent to energy and is frequently translatable into heat energy. More particularly for our study here, heat sometimes appears as an important source of motion, as when the extremely high internal temperature of the Sun shoots forth the mass that we call light at 186,000 miles per second. At other times, motion, and especially repeated motions, present themselves as quite significant causes or sources of heat, as when the repeated tidal massage of the land mass produces sub-surface heating.

The localization of heat in particular areas of earth movement of air, producing Extreme temperatures under the rocky mantle of the Earth cause the earth to erupt into volcanoes, releasing stored up heat far greater than that obtained from burning fossil fuels. Under earth's volcanoes is enough heat that man could discontinue the use of fossil fuels. Technology for use of geothermal heat is in place, and is well advanced. Iceland and California are examples.

Still another important relation between motion or change and heat presents itself wherever there is fire. We witness a forest fire consume the branches and trunks of trees, and send off heat and smoke in a change that could be thought of as a transformation of the elements. Especially important for all that takes place on earth is the movement that takes place within the Sun to which we are linked in more than one way. Whatever matter the Sun encounters in its race through space is taken inside the Sun and at extreme temperature is radiated out as light energy., and completely disintegrated The Sun might be called the great moleculovoir.

In brief, heat and motion are linked. In physics "force X distance" equals work. And when we now consider several major sources of heat on the earth we will need to keep in mind the motions that produce or are linked to these sources of heat. Moreover, we will have to remind ourselves to think of these motions in terms of astronomical time. To come to an intelligible understanding and a credible appreciation of the sources of heat discussed below, some of which might appear novel, we need to hold in mind the fact that nature has been kneading the Earth for billions of years to produce and maintain the life supporting temperatures that we enjoy on this planet. Are there other planets that also support life?

Perhaps that question can be addressed more reasonably after we consider the sources of heat that warm our own planet, where we know there is life.

One major source of heat with which we are quite familiar is the heat from sunlight. Perhaps we are not quite so familiar with the idea that sunlight has weight. It has been estimated that the weight of sunlight that has come to the earth in the last five billion years amounts to 10 to the 17th kilograms. We need to think of the sunlight as heated matter striking the Earth, and in large part retained by the Earth. The light that does not bounce off the Earth, but still remains, must be a part of the mantle of the Earth. If we are able to devise a method of reading the clues that have been left for us, we should be able to determine that the composition of Earth's mantle is congruent with the spectrum of sunlight!

A second major source of heat, already mentioned in the last chapter, springs from the chemical and physical reactions *taking place in the atmosphere* over the North Pole. When we look up at the aurora borealis, that is, the northern lights, we are in reality looking into the firebox of the Earth's heating stove. 4000 miles across and several miles thick. As the

Earth speeds through interstellar space, it collects hydrogen—molecular, atomic, and ionic—from the mostly hydrogen mix of interstellar matter. The hydrogen ions (and a small amount of other positive nuclei) combine with *bio-oxygen* already present in the earth's atmospheric envelope, most probably given off from elementary plant life either in the atmosphere or locked onto the surface of the Earth. (Our oxygen is produced by growing plants!) This interstellar hydrogen can be said to be mostly nascent. When Earth passes through the interstellar soup of hydrogen ions these hydrogen ions rob an oxygen atom from an oxygen molecule. One result of this simple chemical reaction is water. The orphaned oxygen atom joins a bound pair of oxygen atoms (one molecule) to form ozone, a threesome. Both water *and ozone* are necessary for *life* on Earth, as we now know it. The ozone combines with poisonous gases in the upper atmosphere like methane and ammonia, thus purifying the air of poisons.

But equally important as the water and ozone is the heat that is produced. Here some very basic activity, well known to elementary physics, joins the forces active in the chemical reaction. The Earth, in its 200,000 MPH race toward the North Pole, acts as an electrical condenser or capacitor. The condenser is charged by the movement of the Earth and is discharged as lightning. One plate of the condenser is the Earth and the other is the ionosphere. (The Van Allen belt is the bow wave of the earth.) Hence we can say that the Earth is the daddy of all lightning, and the interstellar soup that acts as the fuel for the resulting fire is the mother of all lightening. We note that the northern lights have the spectrum of burning hydrogen plus traces of other elements that add greatly to it's luminosity. In any event, we have here another source of heat that might someday be quantitatively measured. To appreciate the amount of heat produced, we need to think of the heat from a flame perhaps 4000 miles across and 5 or 10 miles thick and dense enough to burn up all the interstellar hydrogen that meets the Earth in its flight. The ozone layer is a *reaction* zone. Predictably, we note the lack of ozone above the South Pole, which is far removed from the production zone. Predictably also, the North Pole is warmer than the South Pole and the rainfall is, on average, *much* more plentiful in the region of the North Pole than around Antarctica. Both the *water* and the *heat* are <u>produced</u> in the north. Human lack of stewardship can indeed affect the ozone layer. But the presence and preservation of the ozone layer depend a thousand times more upon the astronomical activity of the Earth and of the matter through which it travels as captive of the Sun than they depend upon the activities of the s*mall* creatures that live within the Earth's atmosphere.

The third source of heat making life possible is the tidal *massage* of the Earth by the moon and to a lesser extent the massage by the more distant Sun. The two combine more severely once a year when the sun is closest.

The Kobe earthquakes were just *exactly* one year apart.

All heavenly bodies are attracted to each other, and every heavenly body has an effect upon the other heavenly bodies. We refer to this mutual attraction as the gravitational force that unites and holds all the stars, galaxies, and other heavenly bodies together in one cosmos or universe. For our present study, however, it is the attraction beam between the Earth and its moon that chiefly concerns us. We can borrow from Buck Rogers and call these forces between the bodies "tractor beams" and thus benefit from the Latin derivation of the word since "tractor" which in Latin means to *draw* or *pull*. The tractor beam from the Earth to the moon is just strong enough to anchor the moon in orbit about the Earth; It's strength is *exactly* the mass of the moon.. The beam thus prevents the moon from leaving the scene and guarantees the continued influence of the Earth and moon upon each other. The tractor beam is anchored at each end. Imagine that the beam between Earth and moon ends in a big fist at each end, one clasped about the Earth and the other about the moon. Both Earth and moon must work to rotate within a fist. Stated in other words, the gravitational pull of earth and moon on each other tends to hold back the rotation of each on its own axis. One very important result of this work of turning against the gravitational pull is sub-surface heat. (*Volcanoes*)

At some time in the past, the moon became locked, frozen, within the fist from the gravity of the Earth. For this reason the moon keeps always the same face toward the Earth and is said to be in a tidal freeze. In the past, when the moon rotated on its own axis, it was subject internal heating from the tidal force of the Earth, even as Earth still today is subject to the tidal force exercised by the moon. Jupiter's moon Io is the classic example of tidal heating, it is boiling! At one time, in fact, the moon rotated fast enough so that the massage by Earth caused internal heat within the moon great enough to melt much of the rock inside the moon. At some time in the ancient past the moon was so hot that the insides boiled out into huge lava flows, areas much larger than the state of Texas. These areas on the surface of the moon, now no longer molten because of the tidal freeze that now holds the moon in stationary captivity in relation to the Earth, are known as "maria" from the Latin word for seas. Some day reliable quantitative estimates will be made of the size and temperatures of the heating necessary to create these maria.

We know that other planets circling our Sun have natural satellites similar to the Earth's moon. These planets are subject to tidal heating caused by the grip of their "moon". Hence there is possibly present one condition, namely supportive heat, that is necessary for life similar to that on earth. It is estimated that the Milky Way, our home galaxy, is composed of 400 billion stars. *If* each of these stars had as many associated satellites as does our Sun, there might be trillions of chances for life out there. (cloning tells us that life in our galaxy is one big family- how humiliating to mankind, our passage will not be missed) Not all of these moons, however, even those within our solar system are large enough to provide heat similar to that found under the mantle of Earth. Eventually these natural satellites will fall into their captor planets due to the friction. Velocity decreases, even as mass increases; hence the distance diminishes, following Kepler's third law.

To imagine this source of heat, think of the Earth as turning underneath the moon, that is, underneath the gravitational pull of the moon as it grips the Earth. Think that this turning against the grasp of the fist at the end of the tractor beam takes *effort* (which it does) called work. Force times distance equals work. In this instance work is translated in major part into heat. This tidal massage of the Earth by the moon as she circles the Earth each day is a third major source of heat, making life possible on earth. (The force is big, and the same all over, but where there is movement for a long time there forms enough *isolated* heat to form a volcano!

We see one effect of this grip by the moon in the tidal movement of the oceans repeated day by day. Our concern at present, however, is the massage of the mantle of the Earth, and particularly of the mantle that lies beneath the oceans and at the edges of the seas. The tidal force crunches and bends the mass of the Earth; this causes the Earth to get hot. Again, force times distance is equal to work. Here work shows up as heat. The collection of heat results in a rise in the temperature both within the Earth and it is conducted to the atmosphere helping to make us *warm.*

If at first we are not impressed with this source of heat, we need to consider the pockets of molten magma beneath the volcanoes, which has been produced by this tidal massage of the Earth by the moon. We need to remember that the moon has had billions of years to raise the temperature. This tidal force is exerted upon the whole surface of the Earth, but the heat gets concentrated in spots where the mantle is weaker, anyplace the rock will *move*. As these spots get hotter and hotter, they become still weaker. Higher temperatures, in spots, within the earth show where concentrated movement from the force of the moon occurs. When the heat from the crunching action of the moon's tidal force becomes concentrated to such an extent that it bursts through the mantle, a volcano is born.

As noted above, a similar concentration of the force of the Earth tides on the moon, in the past, melted the sub-surface rock of the moon. (Lava flows as big in area as the USA State of Texas)

In summary, there are THREE main sources of heat affecting the raised temperature on the Earth - - sunshine, the heat from the hydrogen-oxygen reaction and the heat from the tidal *massage* by the moon. Evidence for all three of these sources is available either from direct observation or observation of their effects. All three sources are understandable from observable interactions of the Earth with its surroundings. There is no need to propose some huge atomic furnace within the Earth, producing magma. The explanation is extremely simple. The interior of the Earth is heated by the massage of the tides! Something like a microwave. This means that to find another pocket of life like our own in the universe, we must look for a body that has a moon, that is, a natural satellite that can massage its parent and produce heat. And look for the *hydrogen-oxygen* spectrum. Mars once had a moon. There are two fragments left today. We know that Mars once had a moon because Mars has dead volcanoes. When Mars lost its moon, the outermost crust shrank from the cooling and increased the pressure on the pockets of still molten rock within until these erupted, something like a pimple bursting on the surface of a person's face. We also see signs of previous water on Mars, in surface features. It is not a far-fetched notion that Mars once had life, *eons ago*. Another example of this action here on Earth is the water bursting forth when the new frost meets the permafrost in Canada and Siberia each fall.

7. Formation of Mountains, Predicting Volcanoes and Earthquakes

We can speak of three distinct forces that are operative and important in the formation of mountains on Earth. One force pushes southward from the North Pole, due to the centrifugal force exerted by the spin of the earth. Regularly assisted by 10,000 feet thick *ice sheets*. The tidal force from the moon, and to a lesser extent that from the Sun pulls ever westward. And a third subterranean force displays itself in the volcanic release of the extreme heat that has built up under the Earth's mantle.

The first mountains on earth were most likely formed in a ring around the northern, forward-moving nose of the Earth as it sped through space striking and collecting whatever was in its path. In this mountain making work the Earth is an *aggressor* of substantial size and energy. The Earth is 8000 miles in diameter, giving her a frontal area of about 50 million square miles. If the Earth travels toward the North Star at an estimated rate of about 200,000 miles per hour, she sweeps up the material contained in about one trillion cubic miles of space *per hour*. The mass of material collected and retained during this trek that lasted billions of years, was immense. It is the continents *and more.*

To the extent that this mass remained on the surface of the Earth, it became subject to the centrifugal force caused by the spin of the Earth on its north-south axis. As the earth turned, the matter held on its surface slid and was pulled away from the central region about the North Pole and it moved southward. Gradually, over millions of years, this mass grew ever larger, further overcame the frictional force holding it in place, and pushed ever southward. When the forward portion was too much on a slant to slide, the rearmost portions slid up on the forward, causing a "chubbying up "that eventually became the early beginning of a mountain range. An early example is the mountain backbone of North and South America.

At some point in the early history of the earth this mass of matter became stabilized in a ring around the North Pole as far from the Pole as is the Brooks Mountain Range in Alaska today. Over fifty percent of this ring of mountains is still in place, it must have built again and again. In time and surely aided by newly arrived water that formed a huge sea within the mountains about the North Pole this original ring of mountains broke in places and a portion of it swung south to form the mountain backbone of the Western Hemisphere. At some point early in the scheme the landmass that is now known as Australia split off from South America. As noted above, evidence for this split is the fact that the mountains of Australia are located on the eastern edge of that continent rather than on the western edge as they are along the pacific coast of South America. Living fossils; the marsupials in these now separated regions give additional evidence that the ancestors of the possums and kangaroos once shared a common homeland.

A second mountain-forming force is the daily massage of the Earth by the moon and to a lesser extent by the sun that we know as the tides. The tides of the moon tug on the whole surface of the Earth in the daily rotation of the moon about the Earth. The daily rotation of the earth on its own axis similarly distributes the tug of the Sun over the whole surface of the Earth. These two rotations account for the periodicity of the tides that pull on the Earth. The variations in the timing and force of the tides are affected by many factors including the distance between the Earth and moon and Sun, the surface of the Earth affected, the angle of the tug in relation to any portion of the surface, atmospheric conditions, and still other factors.

Even though the tug of the moon is felt over the whole surface of the Earth, the tidal force is applied also to the various parts of the earth, and with differing intensities and effects. The tidal pull on the ocean moves the water daily in a way that is immediately evident, and to some extent measurable. The same tidal force, in some way focused by the oceans, acts as a huge diaphragm tugging at the land beneath the oceans. This massage heats the ocean floor underneath at focal points the surface mantle. The tidal massage is fairly regular, but its effect on various portions of the Earth is not uniform. The heat becomes concentrated at the weakest spots; because of their weakened condition these areas become increasingly more movable to the repeated force applied to them. The rock beneath the surface of the Earth becomes heated to ever greater depths. As the concentrated accumulated heat melts the rock deep below the surface, the force of the massage increases in an auto catalytic manner. This raises the temperature still higher until the pressure of out-gassing becomes so great that the heat breaks out of its *enclosure* within the mantle of the Earth. A volcano is born, spewing fire and molten rock or lava from its *blowhole*. As the molten rock cools and hardens we have the beginning of the formation of a new type of mountain, a *volcanic* mountain. Because the mantle still hot, at the center of the volcano remains weakened after the first eruption, a second eruption, and a third, and more tend to occur in the same place and the volcanic mountain grows.

The mouth of the volcano provides a vent for the gases from the "cooked" rock to escape. In contrast the heat that melted the rock leaves the re-hardened rock away from the mouth more dense, with a higher specific gravity than formerly. Earth has the highest average density of all the planets orbiting the Sun. I believe this means that the Earth is the oldest of all the planets. This fact is an index of the importance of the heat-producing effect of the moon, and its relation to the possibility of life on Earth. It also raises interesting questions, such as how deep a pocket of magma can be, and whether we have not already been given enough information to go about learning how to control the eruptions of volcanoes. Could it be possible, for example, to drill so deeply in the area of an incipient or threatening volcano so as to relieve the pressure? At least in some instances, the earth forms a "bulge" before a first eruption of a volcano, as for example in Palmdale, California.

In the creation of volcanoes we see one connection between the tidal massage by the moon and the formation of mountains. Orbiting bodies in the solar system drift ever closer to their captors. Our moon, too, has been coming closer to the Earth. This means that the effect of the moon upon the Earth becomes ever more intense; and this, in turn, points to the possibility of ever bigger or more frequent volcanoes—The "no free lunch" principle.

A significantly different and quite important connection between the tidal massage by the moon and the formation of mountains can be seen in the movement of Earth's large landmasses by the combined force of the tides. This raises the question of the possibility that the tidal forces might be able to move continents, or parts of continents, that could not be moved earlier. For example, will the splint of land that is Lower California, break loose and move due westward as Japan has done in the past. The sites of Tokyo and Seattle were once neighbors.

As noted earlier this tidal force is always westward because of the direction in which the moon circles the Earth, and upwards, that is, away from the center of the Earth and toward the moon as the source of the force. As also noted earlier, we have to hold in mind that this force has been exerted constantly for billions of years. One mass is pulled up over another in a huge rippling of the surface as the combined tidal forces, working over the space of eons, resist the turning of the earth on its axis. What takes place in this movement of the landmass is not the subduction of one plate below another, but an overduction of one plate upon another because the pull is upwards.

It is time, in fact, to refine and perhaps to change our thinking on the notion of tectonic plates. Instead of thinking of plates we need to think in terms of the "hinge lines" in the subsurface. These "hinge lines" are planes of weakness; and even though cast in stone they do shift with time, and do so repeatedly. The most notable hinge lines are the edges of the Pacific Ocean, where ninety percent of the volcanic and earthquake activity takes place.

Of particular interest in this regard is the chain of Hawaiian Islands. Why do these islands lie in a chain? Why does each younger island become larger than the last? How are we to understand that there are other islands and sea mounts under the waters of the Pacific, that display some of the same tendencies, but which go off in various directions? We know that the moon causes volcanoes. We know that the earth has a 23-degree wobble, which it executes every 17 thousand years. Is each volcanic mount 17 thousand years older than its earlier sibling is? Why is there a string, and why are the islands at particular angles to one another? We need to consider also that the ocean is at some times deeper, or more shallow than at other times, perhaps by as much as a thousand feet. Does this difference in elevation and quantity of water affect the concentration of the moon's huge force on various *focal spots*?

Earthquakes occur when the stored up stress caused by the effort to shift a land mass reaches the point when it must be relieved. The recent earthquakes in the city of Kobe in Japan, being exactly one year apart, were probably triggered by the Sun, which is nearer to Earth once a year and therefore exerts greater tidal force. In other earthquakes, the after-shocks tend to be twenty-four hours apart, suggesting a relation to the tidal forces of the moon.. Earthquakes could be artificially triggered by buried atomic bombs planted nearby, or perhaps by a much smaller trigger. Humanity has the means to relieve, in a controlled fashion, the stress that causes earthquakes and need no longer endure the tragedy of surprise earthquakes. We have been given the evidence; it is up to us to make use of it.

The shifting of the landmasses results not only in volcanoes and earthquakes and in the particular placement of the continents, but also in other interesting formations. Rift valleys are formed by the westward tug of the tidal forces. A good example of this phenomenon is the series of Finger Lakes in upstate New York. *Stretch marks.* During the repeated ice ages that the earth has undergone, the mountain-building force of the tides is supplemented by and diverted by the push of the ice. To get an idea of the force of an ice age we might think of a sheet of ice perhaps two miles thick and six thousand miles across,—pushing. There were four ice ages in the last quarter of a million years. If we can extrapolate from this brief record, there should have been 1600 ice ages in the last 100 million years. The largest of these ice ages are recorded in the Himalayan mountain ranges to the north of India. Each big push is recorded in a ridge.

8. Related Issues for Further Study

When we begin to read the heavens with an eye and mind less bound by conventional ways of thinking, issues that once seemed fairly settled or seemed beyond the realm of possibility invite us to think anew and to look again for evidence in new places or in new ways.

The importance that the account above places on the massage of the Earth by the moon, for example, is not limited to the question of heat on Earth. It teaches us also that the question of life elsewhere in the galaxy is likely related to the presence, the size and nature of planetary satellites. It suggests also that there may be other important relations, not yet discovered, between the planets and their lesser companions.

The gathering activity of the Earth and other bodies as they race through space might already contain evidence, if we search for it, that one very significant element of this gathered material is DNA. What kinds of equipment do we need to develop in order to read the evidence that is waiting here to be read? Are the regions that hold out the greatest promise of evidence the areas of space behind the Earth and other bodies, that is, the *tails* that have been grown by these bodies? Going beyond the question of the presence of DNA can we not think of presence and of changes in DNA in the vast regions of space and in the vast periods of astronomical time? Can we think of the atmosphere of the Earth as a huge Petri dish? If so, cloning has been going on for billions of years. And the question of evolution is not *earthbound.*

On a question that is more earthbound, we even now have sufficient evidence about the causes of earthquakes to be able to predict their arrival with some assurance. Have we already been given enough evidence, if we can pool and develop it, to be able to control the devastation of earthquakes? This might be done by some artificial triggering that releases the pressure below the mantle of the Earth before it becomes so excessive that we can only pick up the pieces. A similar question needs to be asked regarding volcanoes. Is there sufficient evidence for us to begin to learn ways to measure the heat in a magma bed and to cool it? If we could learn how to harness the heat of volcanoes and the force of earthquakes, we could tap into a source of energy much greater than we enjoy now in fossil fuels?.

Another issue related to the study of astronomy in its relation to life on Earth, is the issue of the broader meaning of the evidence that is available to us. Evidence is generously available to us in every aspect and in every location of the cosmos. The Maker has left his mark in the billions of stars, in the rigidity of stone and mountain, in all the movements of the bodies large and minute, in the vast regions of what we perhaps falsely think of as empty space, and in length of days and eons that boggle our minds. Our vision is indeed myopic if we cannot see this evidence as the care that the Maker takes not just for mankind alone, but for all of his creatures. What then is the place of man in the cosmos?

About the Author

James H. Pax is a WWII veteran who was born in New Weston, Ohio. He earned a BS at the University of Washington in 1948. He worked in manufacturing for sixty years.